Penguin Coloring Book for Adults

Penguin Coloring Book containing Penguins filled with intricate and stress relieving patterns

by The Coloring Book People

ISBN-13: 978-1532714429

ISBN-10: 1532714424

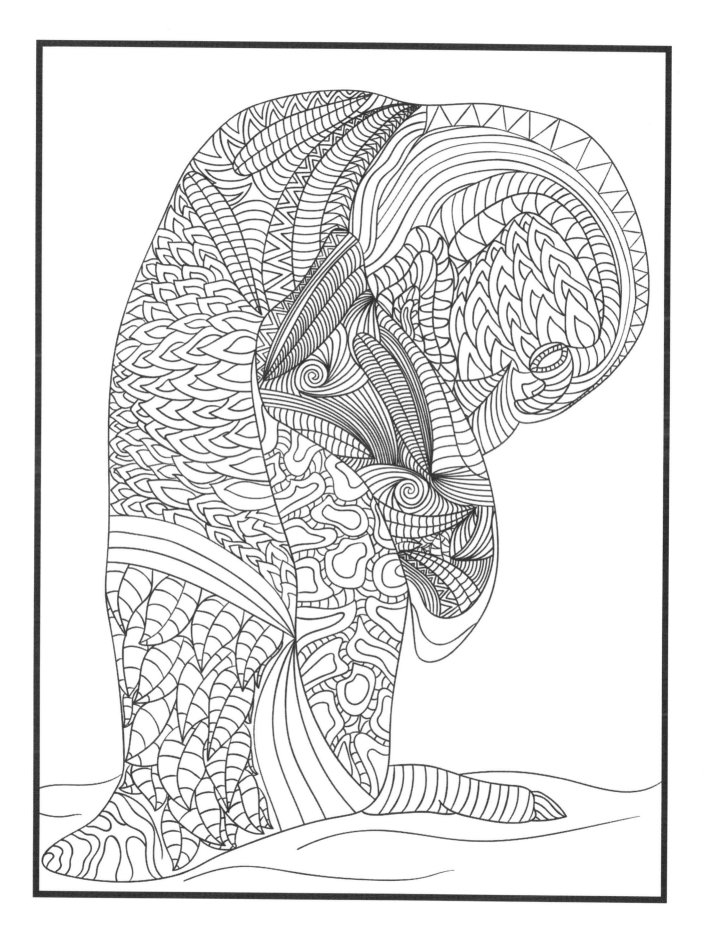

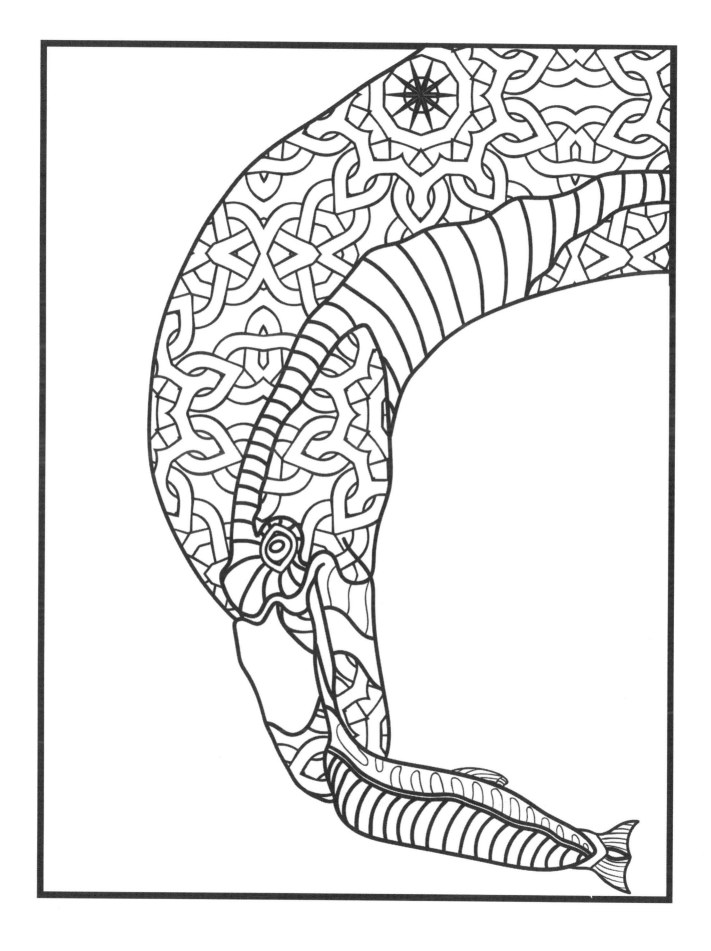

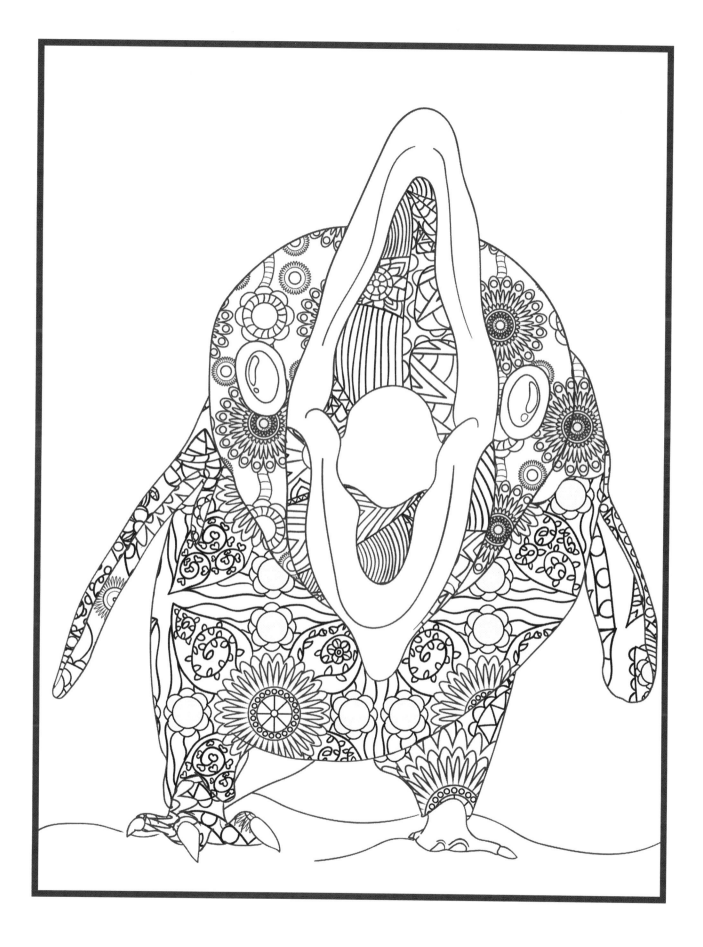

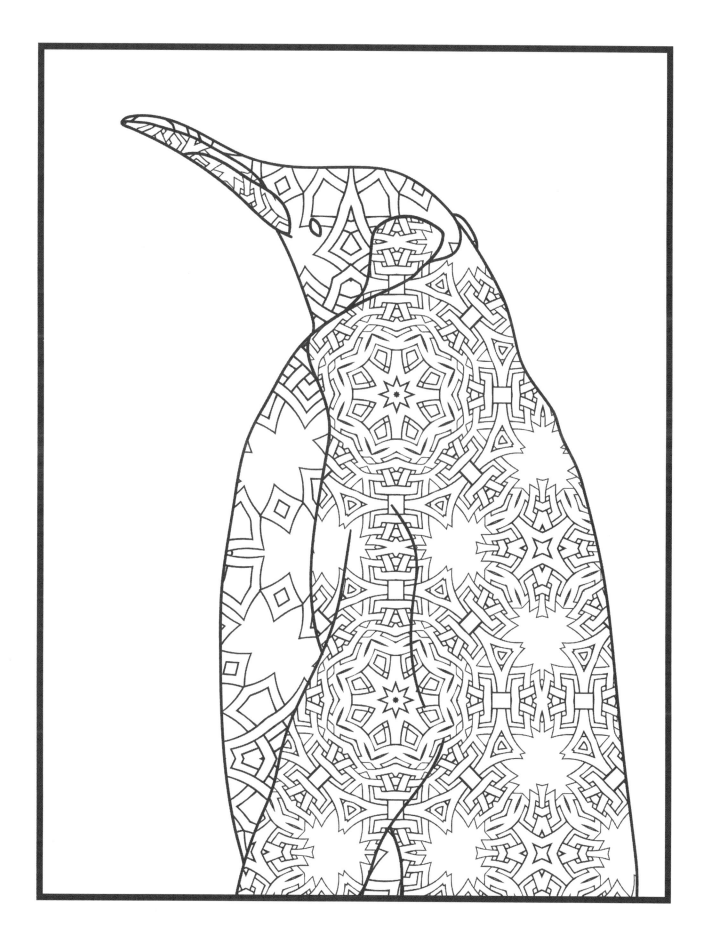

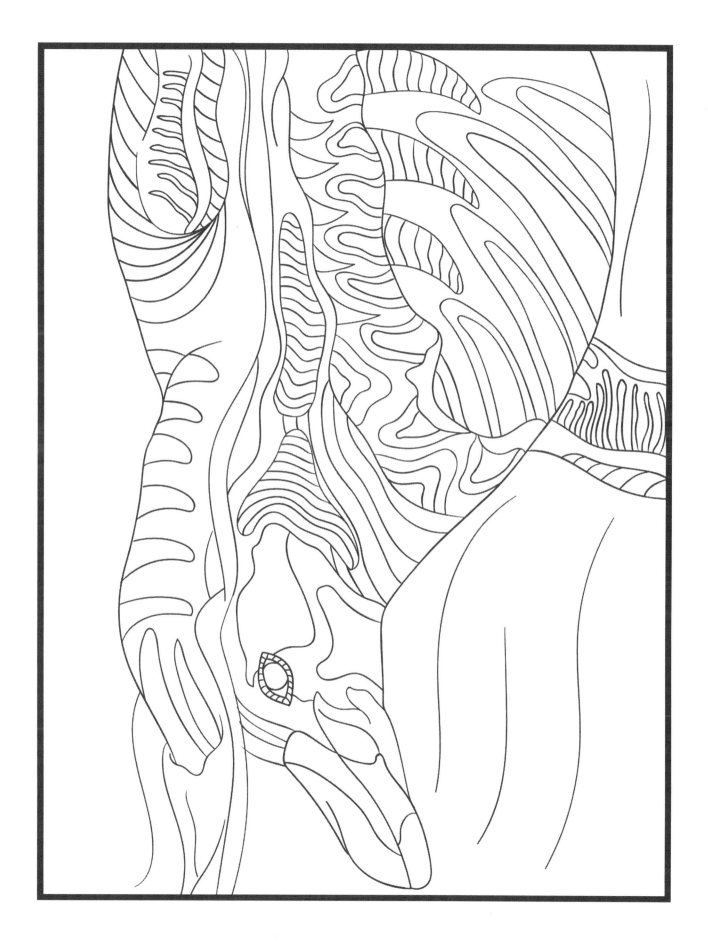

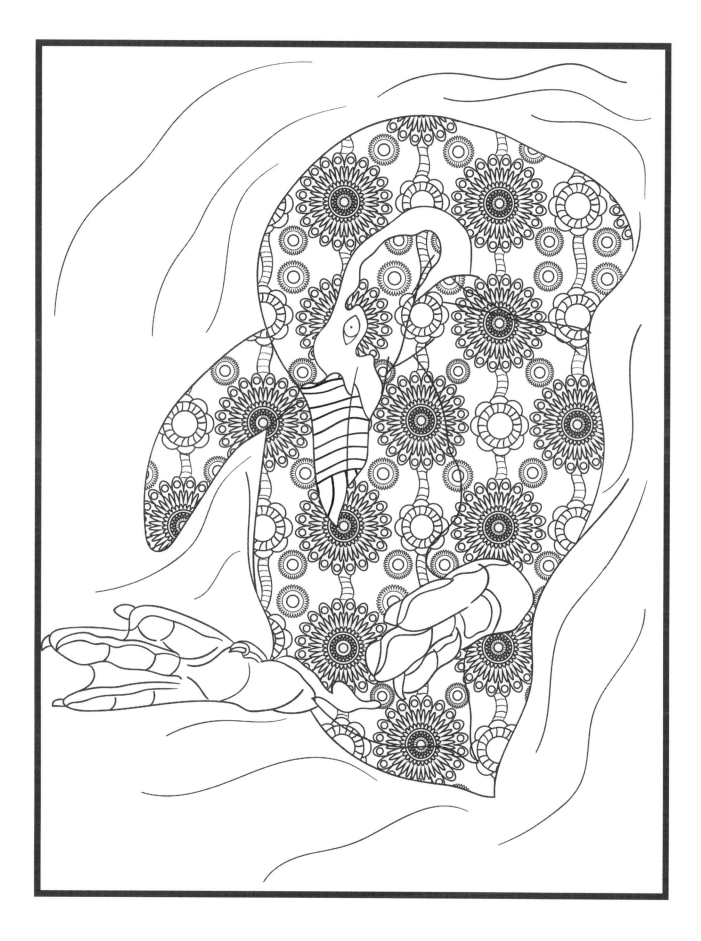

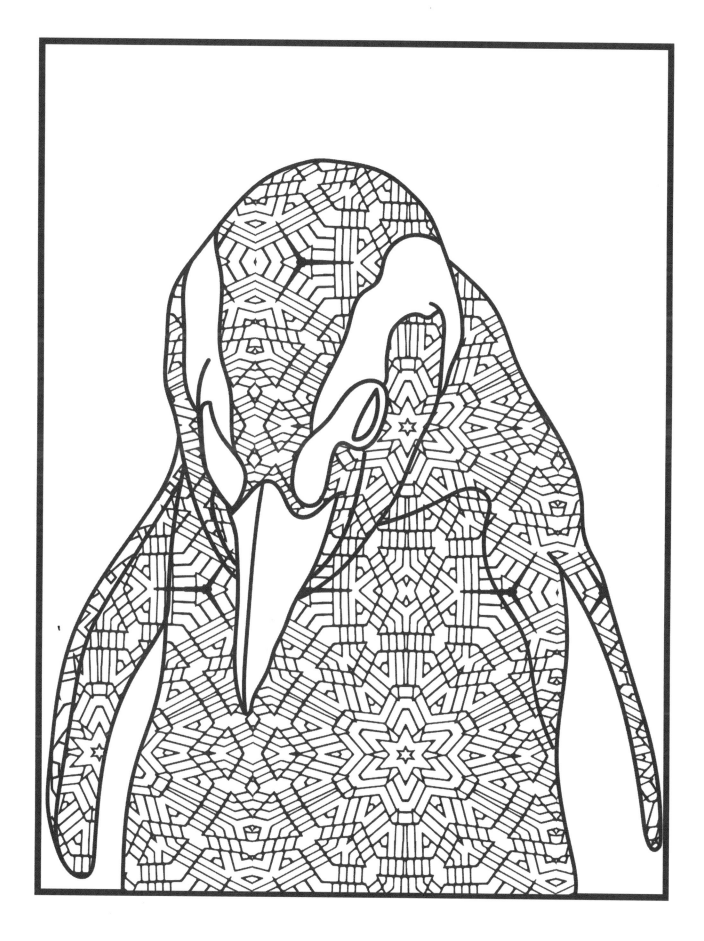

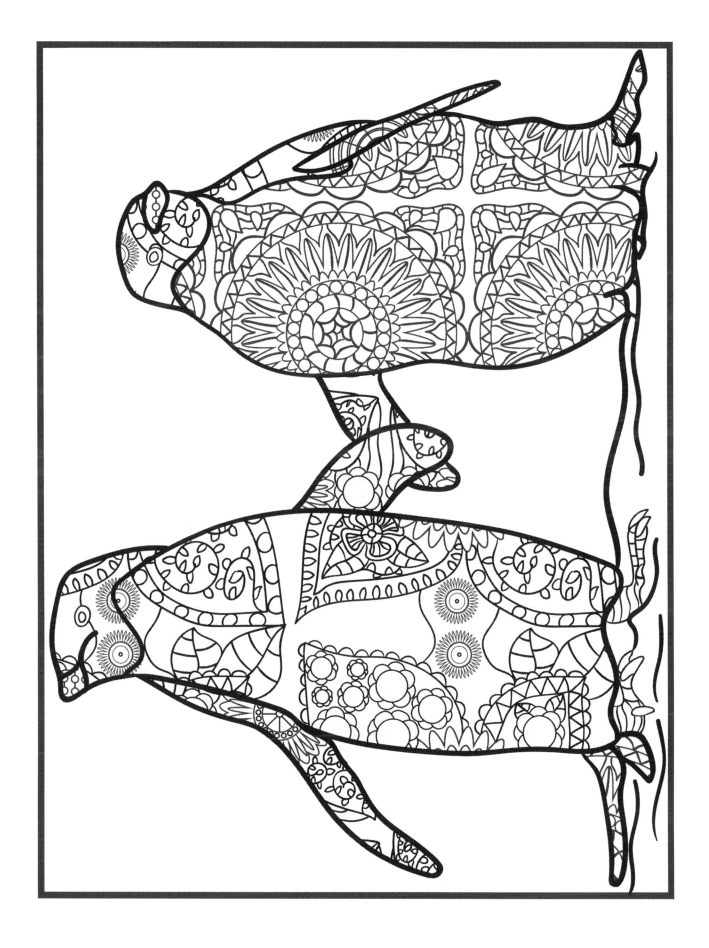

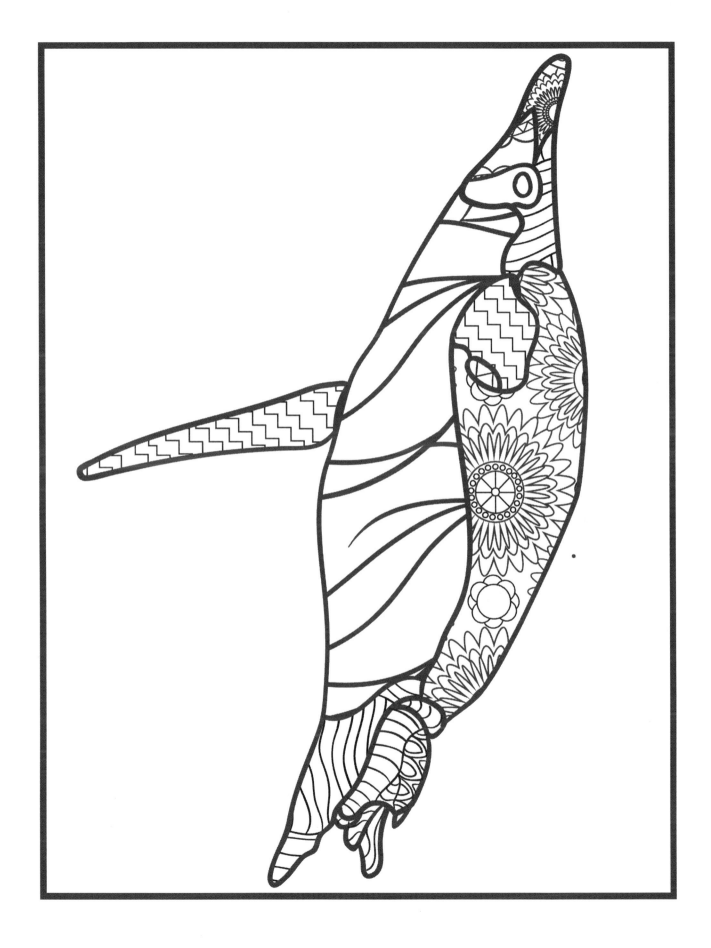

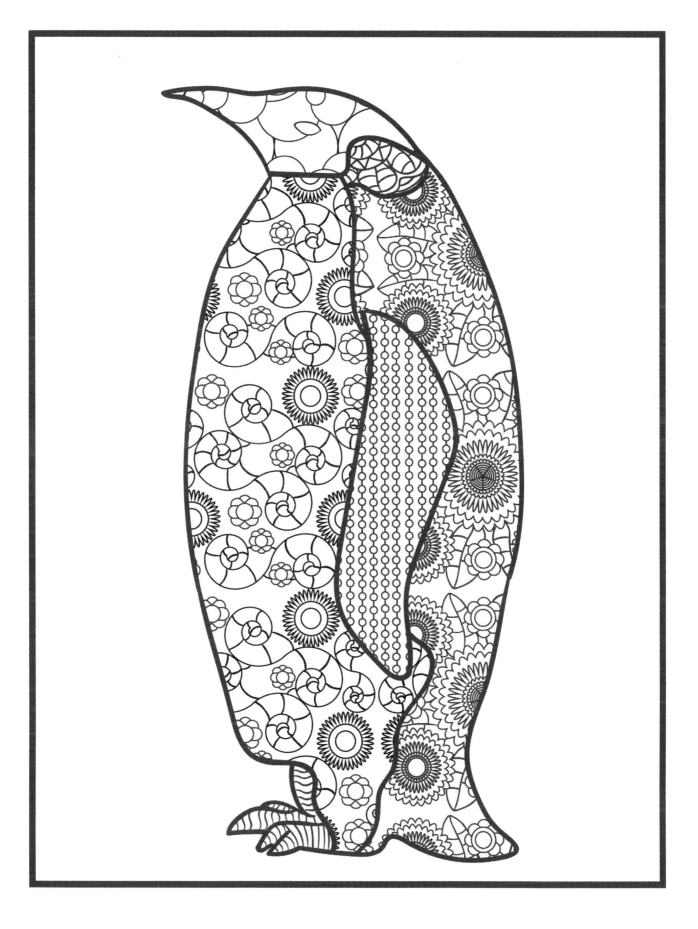

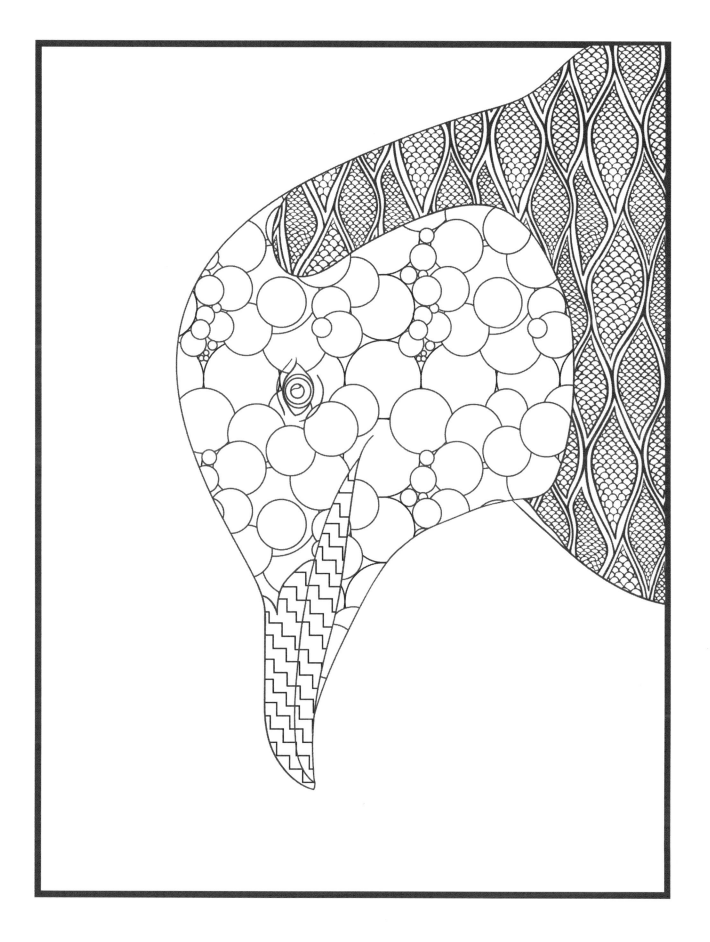

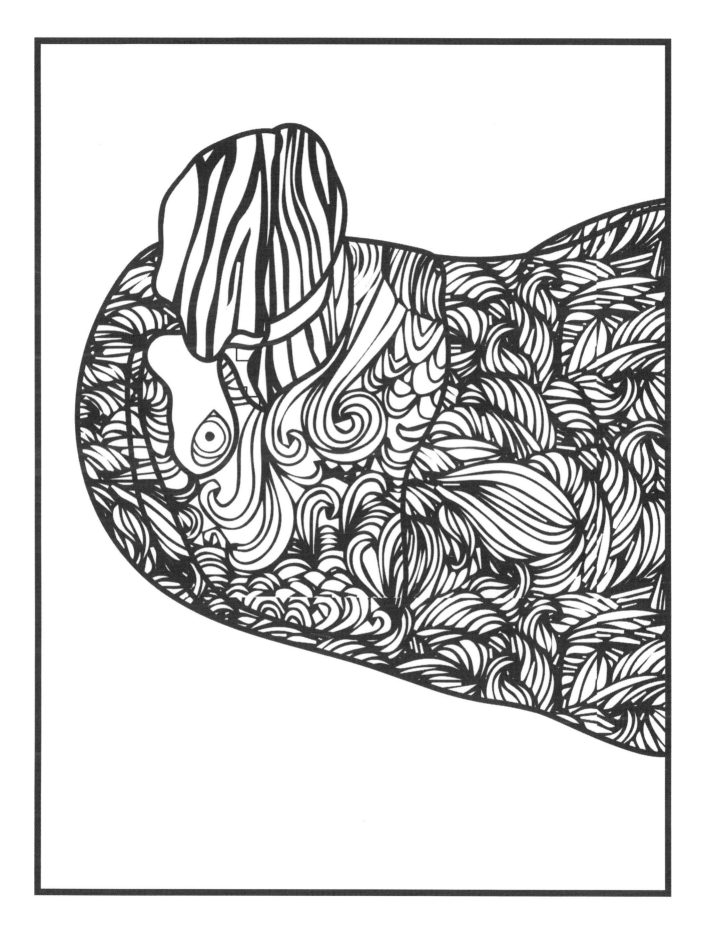

COLOR TEST PAGE

COLOR TEST PAGE

Made in the USA
Lexington, KY
01 April 2017